YOUR KNOWLEDGE HAS VALUE

- We will publish your bachelor's and
  master's thesis, essays and papers

- Your own eBook and book -
  sold worldwide in all relevant shops

- Earn money with each sale

Upload your text at www.GRIN.com
and publish for free

Emmanuel Boon, Albert Ahenkan

Enhancing food security and poverty reduction in Ghana through non-timber forest products farming: Case study of Sefwi Wiawso District

GRIN Verlag

Bibliografische Information der Deutschen Nationalbibliothek:

Die Deutsche Bibliothek verzeichnet diese Publikation in der Deutschen National-
bibliografie; detaillierte bibliografische Daten sind im Internet über http://dnb.d-
nb.de/ abrufbar.

Imprint:

Copyright © 2008 GRIN Verlag GmbH
Druck und Bindung: Books on Demand GmbH, Norderstedt Germany
ISBN: 978-3-640-14306-1

Enhancing food security and poverty reduction in Ghana through non-timber forest products farming:

Case study of Sefwi Wiawso District

Emmanuel K. Boon

Human Ecology Department, Vrije Universiteit Brussel

Albert Ahenkan

Human Ecology Department, Vrije Universiteit Brussel

Table of Contents

Abstract

Food insecurity and poverty are the greatest global challenges facing the world today. Their redress is an indispensable requirement for sustainable development in developing countries, particularly in Africa. Poverty continues to be pervasive, intractable and inexcusable in the developing world particularly Sub-Saharan Africa. Extreme poverty ravages the lives of one in every four in the developing world (i.e. 1.2 billion people). The farming of non-timber forest products (NTFPs) is being promoted as a potential solution to the current high rates of malnutrition and poor health of the rural population, the degradation of tropical forests and the spread and intensification of poverty. However, the role of NTFPs in rural development in Ghana remains largely undervalued and understudied. This paper examines the contribution of NTFPs farming in enhancing poverty reduction, food security, sustainable forest management, and livelihoods improvement in the Sefwi Wiawso District (SWD) of the Western Region of Ghana. An exhaustive literature review and analysis of field data collected through administration of questionnaires, interviews and stakeholder consultations indicate that NTFPs are an important tool for addressing poverty amongst marginalised, forest-dependant communities in Ghana. 86% of farmers in SWD depend on NTFPs for income, food and medicine. The results of the study support the call on decision makers and development actors to put NTFPs management on national, regional and district development agendas.

Key Words: Biodiversity, conservation, forest reserves, food security, livelihoods improvement, non-timber forest products, malnutrition, poverty reduction, sustainable forest management.

# 1.	Introduction

Poverty eradication and food security have moved to the centre stage of the global development agenda. They are the greatest global challenges facing the world today. They are indispensable requirements for sustainable development, particularly for developing countries (WSSD, 2002).The world's leaders pledged their commitment to eliminate hunger, malnutrition and to halt global poverty by 2015 at the World Food Summit in 1996 and the Millennium Summit in 2006 (WFS, 1996). At the World Summit on Sustainable Development (WSSD) held in Johannesburg in 2002, the international community reaffirmed its commitment to develop national and local programmes for sustainable development, poverty eradication and food security was agreed upon. Despite these commitments, the last decade has witness an increased poverty level, especially in developing and transition countries. (OECD, 2001).

Although the proportion of people living in extreme poverty in developing and transition countries fell slightly over the past decade from (42% to 28%), the numbers of poor people steadily increased in most regions except for East Asia (Boon, E.K, 2005). Ten years after the World Food Summit (WFS), the number of undernourished people in the world remains stubbornly high (FAO 2006). Worldwide, 854 million people still remain hungry and undernourished. 820 million of this figure is found in the developing countries (FAO, 2006; USDA, 2007). Today, extreme poverty ravages the lives of one in every four in the developing world (Boon, 2004). Of the estimated 20 per cent of the world's population (i.e., 1.2 billion people) trapped in severe poverty with less than US$1 per day, 26 per cent are in sub-Saharan Africa (Sackey, 2005).

Clearly, poverty continues to be pervasive, intractable and inexcusable in the developing world, especially in sub-Saharan Africa (Boon E.K, 2004). Poverty and hunger are inseparable in developing countries and are the most common triggers for the downward slide in poverty. Poverty is a major cause of food insecurity and sustainable progress in poverty eradication is critical for improving access to food (FAO, 2006). A food paradox exists in contemporary world. Enough food is produced world-wide for satisfying every person. However, while poor health in developing countries is often linked to inadequate food and chronic disease in the rich industrialized countries it is tied to too much food.

2.   Food Security and Poverty

Food security is a multi-faceted and flexible concept as is reflected by the many attempts at defining and interpreting it by different organizations and individuals (Maxwell et al, 1992). The most recent definition of food security is that negotiated in the process of international consultation leading to the World Food Summit (WFS) in November 1996 which defines food security as "a situation that exists when all people, at all times, have physical, social and economic access to sufficient, safe and nutritious food that meets their dietary needs and food preferences for an active and healthy life" (WFS, 1996; FAO, 2002).

2.1   What is Poverty?

It is generally agreed that there is no universally accepted definition of poverty (Boon E.K, 2005). However, several attempts have been made by different authors and international institutions to find an "acceptable norm" for defining poverty. It is expressed in terms of the income that a household or individual would require to purchase goods and services deemed necessary to sustain his or her physical and social existence. The World Bank Development Report (1990) used $370 per person per year (or US$1 a day) as the absolute poverty line in developing countries and estimated that 1,116 billion people or 33 percent of the population in the developing world are poor (World Bank, 1990).

2.2   OECD's View on Poverty

The OECD (2001) describes poverty as an unacceptable human deprivation in terms of economic opportunity, education, health and nutrition, lack of empowerment and security. In general, poverty is the inability of people to meet economic, social and other standards of well-being. It covers measures of absolute poverty such as child and infant mortality rates, and relative poverty as defined by the differing standards of each society. Figure 1 depicts a broader framework of the concept of poverty developed by the OECD. It includes all the most important areas in which people of either gender are deprived and perceived as incapacitated in different societies and local contexts. The framework encompasses the casual links between the core dimensions of poverty and the central importance of gender and environmentally sustainable development.

Figure 1: Interactive Dimensions of Poverty and Well-being

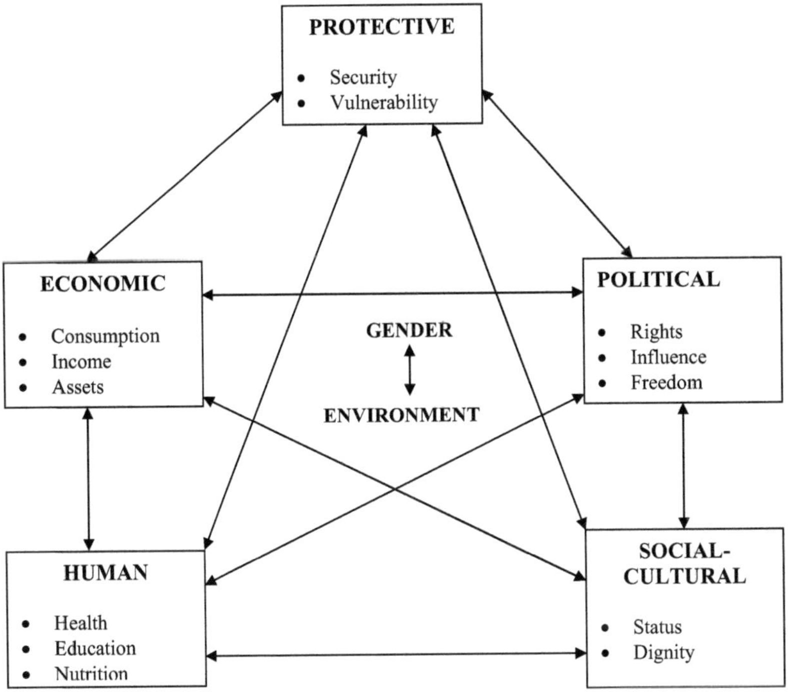

Source: OECD, 2001

Poverty is a multi-dimensional and dynamic construct. The dimensions of poverty can be categorized into five main facets: economic, human, social-cultural, political and protective. The economic dimension of poverty implies ability to earn income, levels of income or low levels of consumption that are socially unacceptable and ability to have assets which are all key to food security, material well-being and social status. The human dimension of poverty includes lack of access to health-care, education, good drinking water, decent housing, and healthy sanitation meant to improve livelihood (Boon E.K., 2005). The socio-cultural dimension includes the ability to participate as a valued member of a community. The political dimension includes lack of voice and political rights. People who lack the ability to participate in decisions that affect their lives directly consider this as a sense of helplessness and a fundamental characteristic of poverty. The protective dimension includes the ability of the people to withstand external shocks such as hunger and food insecurity, natural disasters, economic crises etc.

2.3 Poverty Trends in the World

The World Bank estimates that 1.2 billion people live in extreme poverty (World Bank, 2004). This number is growing steadily as civil wars, loss of employment and restructuring of societies are creating newly poor groups (UNESCO, 2002). Extreme poverty ravages the lives of one in every four in the developing world. Of the estimated 20 per cent of the world's population (i.e., 1.2 billion people) trapped in severe poverty with less than US$1 per day, 26 per cent were in sub-Saharan Africa (World Bank, 2002).

Figure 2: Trends of World Poverty (1981-2001)

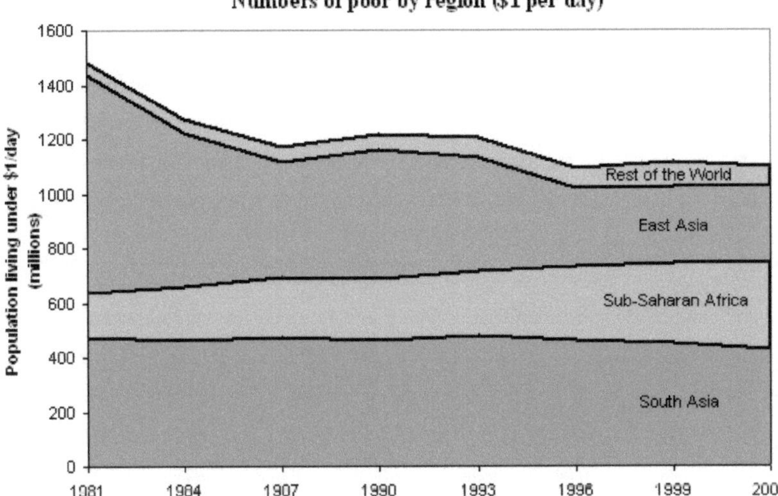

Source: World Development Indicators 2004

It is clear from Figure 2 that poverty in sub-Saharan Africa is expected to remain crushingly high in contrast to progress made in East Asia and South Asia. Although Asia leads in terms of the numbers of poor people, Sub-Saharan Africa has the largest proportion; nearly half of its population lives in extreme poverty. In fact Africa is the only region in the world where the number of poor people has doubled in the last twenty years. While the situation with respect to poverty reduction has gotten better in the rest of the World, in Africa it has gotten worse! The plight of Africa has been described by Carol Bellamy, head of UNICEF, as the perfect storm of human deprivation, one that brings together impoverishment, the AIDS pandemic and the long standing burdens of malaria, and other diseases (Ofosu-Appiah, 2005).

3. Poverty and Food Insecurity in Ghana

Poverty in Ghana is both urban (16.7%) and rural (83.3%) phenomenon concentrated among food-crop farmers (60%) and the rural Savannah where over half the population are extremely poor (Boon E.K., 2005). According to the Ghana Poverty Reduction strategy Paper (PRSP) published in March 2000 by the Ministry of Finance, poverty is widespread in the country, with 42.6% of the population living below the poverty line. Though absolute poverty continued to decline in the 1990s (from 36% in 1992 to 29% in 1999), the declines were not uniform across the country Poverty is widespread in the country and needs to be addressed in all regions. Extreme poverty rose in the savannah and rural coastal regions and 58% of those identified as poor are among households for which food crop production is the main activity (Asante, 2004). Table 1 and Figure 3 depict the regional variation of poverty levels in Ghana. The Upper East, Upper West and Northern Regions are the worst hit, with between 7 and 9 people out of 10 being poor.

Although agriculture is still the mainstay of the economy of Ghana, most farmers are still living in hunger, abject poverty and poor livelihood conditions because of the inability of the country to produce sufficient food to feed the population. For instance, in 2002 the overall domestic production of food in Ghana was deficit. The country was only 63% self-sufficient in cereals production, 60% in fish production, 50% in meat production and less than 30% in raw materials for agro-based industries (Asante, 2003). Food and cash crop agriculture is predominantly rain-fed. Food production fluctuates, both quantitatively and qualitatively, from year to year due to frequent occurrence of climatic stresses during the growing seasons. Sometimes fluctuations of heavy rains within and between agricultural seasons destroy both crops and livestock. Rainfall, the control of which is beyond the capability of the small scale farmer, is thus a major determinant of the annual fluctuations of total household and national food output. This situation creates food insecurity at household levels which can be transitory in poverty areas and chronic in extreme poverty areas.

Agriculture in Ghana is predominantly on a smallholder basis on family–operated farms using rudimentary technologies to produce about 80% of the total agricultural production. About 90% of farm holdings are less than 2 hectares in size. Bad road network linking urban and the rural areas creates a situation of rural glut and urban scarcities in food. It is also estimated that about 20-30% of production are lost due to the poor traditional post harvest management of food crops (MOFA, 2000). Also, growing urbanization has created slums in the cities where unemployment and low incomes appear to be the main constraint to increased calorie consumption. This state of affairs keeps worsening with the years and the nutritional status of city immigrants keeps deteriorating each year. Access to food is determined by a combination of income levels, its distribution and the purchasing power of the population.

Table 1: Regional Distribution of Poverty in Ghana

Region	Poverty Incidence
Upper East	9 out of 10 people are poor
Upper West	8 out of 10 people are poor
Northern	7 out of 10 people are poor
Central	6 out of 10 people are poor
Western	5 out of 10 people are poor
Brong Ahafo	4 out of 10 people are poor
Asante Region	4 out of 10 people are poor
Eastern	3 out of 10 people are poor
Volta Region	3 out of 10 people are poor
Greater Accra	2 out of 10 people are poor

Source: Ghana Statistical Service 2000

3.1   The Link between Poverty and the Environment

The poverty-environment nexus was stressed by the World Commission on Environment and Development. In fact, environment and poverty are linked in many ways. Environmental degradation, in both rural and urban areas, affects poor people the most. Conversely, it is also the result of poverty (Boon, E.K., 2003). In Ghana, farmers have been pushed or squeezed out of high-potential land. Due to population pressure on marginal farm lands, the rural poor often have no choice but to over-exploit the marginal resources available to them through low-input, low-productivity agricultural practices such as overgrazing, soil-mining and deforestation, with consequent land degradation. The cost of environmental degradation in Ghana is estimated to be 5.5 of GDP (Nelson, W., 2007). Environmental opportunities and challenges must be addressed if Poverty Reduction Strategy Paper (PRSPs) are to be effective in eliminating poverty and forging improved livelihoods and sustainable development. As is illustrated in Figure 4, attempts at achieving this goal should be anchored on the inter-relationships amongst the various components of the MDGs. A key premise of this paper is that an effective development and promotion of non-timber forest products (NTFPs) in Ghana will help to significantly reinforce the positive inter-relationships amongst the various components of the MDGs and facilitate food security, poverty reduction and sustainable development.

Figure 4:MDGS INTERRELATIONSHIPS

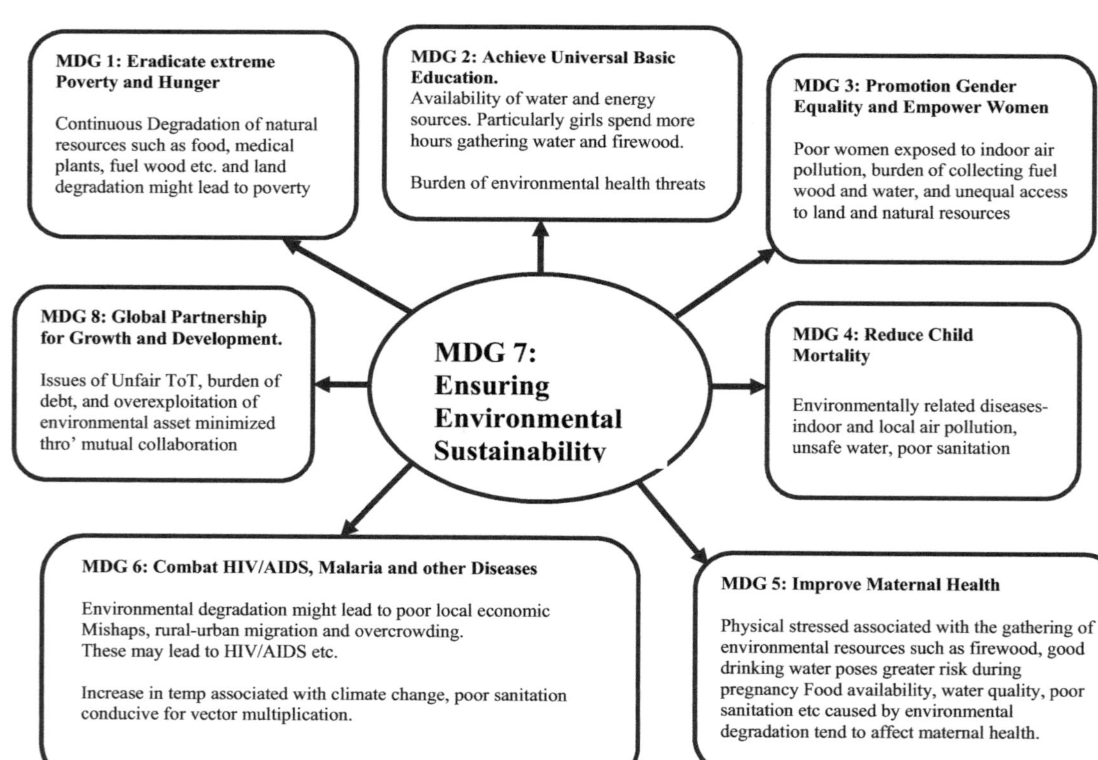

Nelson, W., 2007

4. Non-Timber Forest Products (NTFPs) Farming

4.1 *What are NTFPs?*

The Food and Agriculture Organisation (FAO) was one of the first agencies to promote NTFPs through their programme on *non-wood forest products.* Over the past 10 years, other international agencies such as the World Bank, Canadian International Development Agency (CIDA), International Development Research Centre (IDRC), Center for International Forestry Research (CIFOR), International Union for the Conservation of Nature (IUCN), and the Biodiversity Support Programme (BSP), among others, have incorporated the concept of NTFP into their programming.

Defining NTFPs is difficult because of the *blurred boundaries between timber and non-timber products as well as the underlying difficulty in defining a forest.* Consequently, the term "NTFPs" has generated a lot of *controversy* over its meaning (Shiva et al, 2002; Chandrasekharan, 1995). The debate has raged over its definition since the term was coined in 1989 (Belcher et al, 2005, Neumann and Hirsch, 2000). Currently, there is no universally accepted definition of NTFPs. However, several attempts have been made by different authors and international institutions to find an "acceptable norm" for defining NTFPs (Shiva et al, 2002). An NTFP is defined not by what it is but by what it is not. They are also referred to as "special forest products" (Thomas and Schumann 1993), non-wood forest products (FAO,1997; FAO, 1995), or minor forest products (Malhotra and Poffenberger 1989)

NTFPs are defined as "biological resources of plant and animal origin, harvested from natural forests, plantations, wooded land and trees outside forests" (FAO 1999, Marshall 2005, Peters, 1996). What distinguishes NTFPs from agricultural products is the wild or semi-domesticated mode of production (FAO 1999, Marshall 2005, Shiva et al 2002, Fox, 1994, Peters, 1996). NTFPs are economically, socially and environmentally important worldwide. Examples of NTFPs are fruits and berries, nuts, spices, medicinal extracts, oils, gums, resins, insect and animal products, charcoal, cones, seeds, smokewood and flavourwood, greenery and other floral products, honey, mushrooms, specialty wood products, syrup, weaving and dying materials and aromatics (Thomas and Schumann, 1993; FAO 1995, Shiva et al, 2002). Products such as these are vital sources of income, nutrition and sustenance for many forest-based communities around the world. The commercial development of NTFPs could increase the value of forest resources and thereby reduce pressure on the forest (Marshall et al 2003; Rose-Tonen 1999; FAO, 1991).

4.2   NTFPs and Poverty Reduction

There is no doubt that NTFPs play a significant and often critical role in providing subsistence and *cash income* to a large part of the world's population (Pimentel *et al.* 1997; Marshall, 2005). Studies from all tropical regions indicate that it is often the poorest households in rural communities that are most directly dependent on NTFPs (Falconer, 1992; Marshall 2005, Shiva et al., 2002).This potential contribution of NTFPs to poverty reduction and livelihoods improvement has too long gone unnoticed by development community (FAO, 1991; Marshall et al., 2003).

The current debate over the concept of NTFP depend on the interests and priorities of the proponents, and are usually centred on the *expected contribution of NTFPs to poverty reduction, health, conservation as well as on their current and potential benefits to the poor communities versus their further impoverishment* (Peters, 1996; FAO, 1995; Falconer, 1994). Although development and conservation circles have been interested in NTFPs for decades, there are a number of reasons for the general spread and upsurge of interest amongst them since the 1980s. There is leading to the belief that the promotion of a sustainable use of NTFPs could lead to a *win-win situation for poverty reduction and biodiversity conservation* (FAO 1995; Shiva et al, 2002).

The gathering of NTFPs is a significant *source of employment* for the economically marginalised and forest-dependent communities. Worldwide, an estimated *350 million people depend on forests as their primary source of income, food, nutrition, and medicine* (UNDP, 2004; FOA, 2005). Cash income from the sale of NTFPs such as honey, nuts, fruits, bush meat, medicinal plants, and insect products play a vital role in sustaining the lives of local farmers and gatherers, who must increasingly adapt to diminishing poor harvest and resources to stay alive (Marshall, 2005; Peters, 1996). These resources are critical, especially for the rural poor and particularly the women.

NTFPs may provide them with the only source of personal income (Falconer, 1997 Bhattarai *et al.*, 2003; Edwards, 1996; Andel, 2000).

Achieving the ***poverty reduction and nutrition-related*** goals in developing countries therefore requires that national and sectoral development policies and programmes are complemented by effective community-based actions aimed at improving ***household food security*** and promoting the ***year-round farming, consumption and marketing of NTFPs***. These actions need to occur within the framework of promoting sustainable livelihoods and should address the variety of locally-relevant causes of various forms of malnutrition, including problems of chronic and seasonal food shortages, lack of dietary diversity, inadequate family care and feeding practices, and poor living conditions (FAO, 2005). Also, employment potential related to NTFPs, through trade, harvesting and processing creates a small number of jobs which are often taken up by those with no access to land. People living in forest environments and practising hunting, collecting and agriculture (shifting cultivation) draw heavily on forest products, not only for subsistence but also for income from forest products (Arnold et al., 1999)

4.3   The Need for NTFPs Farming

As a result of the recognition that the ***extraction of NTFPs*** from natural forests has limited potential for improving household economies, several scholars have questioned whether the objective of ***enhancing forest-based livelihoods*** could not be better fulfilled by optimising NTFP production through domestication (De Jong et al., 1998; Arnold, J.E.M. and Ruiz Pérez, 2001). It is often assumed that NTFPs can be sustainably harvested and that this ***"green social security"*** will always be available to resource users. This is not always the case. In many parts of the world, local people are losing access to valued plant and animal species, either through overexploitation and habitat destruction or loss of access as former harvesting areas are included within national parks or forest reserves (Peters, 1994). Ros Tonen et al. (1999) suggest that it is incorrect to think that NTFPs can be harvested indefinitely without proper management practices and domestication to sustain their yield. They therefore call for the need to farm NTFPs.

5. Case Study of Sefwi Wiawso District in the Western Region of Ghana

5.1 The Study Location

The Sefwi Wiawso District (SWD) is located in the Western Region (See Figures 3 and 5) and is one of the eleven administrative and political authorities that form the Western Region of Ghana. It covers an area of 2,166.22 Km^2 comprising of 1,303.12 km^2 off reserved and 863.10 km^2 on reserved forests. The district is located between latitude 4°50' N 5°40' and longitudes 1°45'W and 2°10' (SWDA, 2002). The district forms part of the country's dissected plateau.. The land rises from about 240m to about 300m above sea level. Climatically, the district forms part of the wet semi-equatorial climatic with rainfall peaks. The wet season is between May-June and September-October respectively. It has a fairly uniform temperature ranging between 26°C in August and 30° in March. Relative humidity is generally high throughout the year ranging between 70-80%.

The vegetation consists of the Celtic triplochiton association where some of the trees in the upper and middle layers of the forest shed their leaves usually in the dry season. Large sections of the forest are now secondary type due to farming, logging and bush burning. There is a high degree of depletion of the original forest. Because of this, large sections of the forest (953 km^2) have been put under reserves. It is estimated that over 85% of the off-reserved forests and about 30% of the on-reserved forests have been respectively depleted (Akrasi, 2003). The population of Sefwi Wiawso District is estimated at 148,950 (Ghana Statistical Service 2000) with an annual growth rate of 2.9%. Population density is estimated at 99 per square kilometer with relatively high values along the main roads. This indicates that there is pressure on land. Furthermore, the increase in density is due to the in-migration of people in search of employment in the district (SWDA, 2002).

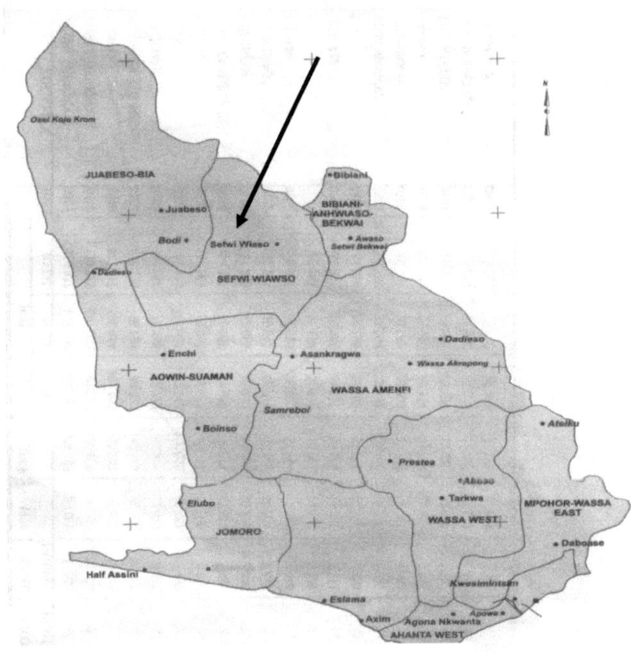

Source: Ghana Statistical Service, 2002

5.2 Impacts of NTFPs Harvesting in Sefwi Wiawso District

NTFPs have generally been considered as free commodities in Ghana and the Sefwi Wiawso District in particular. For most of the rural poor without land or cocoa in the district, harvesting of wild resources is a common option (Akrasi, 2003). However, the present practice of gathering these products from the forest in the district is obviously unsustainable (Boon E.K, 2003). Ros Tonen (1999) stated that it was incorrect to suggest that NTFPs can be harvested indefinitely without proper management practices and domestication to sustain their yield and therefore called for the need for Non-timber Forest Resources (NTFR) farming.

The heavy dependence of the population of the district on NTFPs and the rapid population growth has contributed to accelerate the destruction of the forests in the area.

19

Deforestation and environmental degradation are particularly severe around the bigger villages. In the off reserves, the natural vegetation cover has given way to secondary forest over most of the district because of over exploitation of timber and Non-timber forest resources (Akrasi, 2003).

To check further degradation of the forests, the Government has since 1990 intervened with "Operation Halt" which is purely an enforcement measure. This measure is proving effective in other areas but has unfortunately been met with some stiff resistance from communities in the district. This is because for most of the people without land or cocoa, harvesting of wild resources is a common option for livelihood. Even those having cocoa farms are forced to harvest NTFPs from the forests to supplement their incomes due to the poor performance of cocoa in the district (Boon, 2003). A recent survey in the forest zone in southern Ghana indicated that 10 percent of rural population gains some income, food and medicine from non-timber forest products (FAO 2002; Arnold and Townson, 1998). Aggregate employment in forest product activities has been estimated to be growing at 6.9 percent per year (FAO, 2002). The assessment of NTFPs in Ghana is relatively new and has received very little formal study. To date, no formal studies have been carried out on the long-term impact of NTFPs farming in promoting poverty reduction, sustainable forest management and human development in the communities surrounding them.

5.3 NTFPs Farming in Ghana

In line with Ghana Government's policy on sustainable management of forest resources, biodiversity conservation, accelerated poverty reduction and food security in the country, the International Centre for Enterprise and Sustainable Development (ICED) and its partners are implementing a project on non-timber resource (NTFR) farming in the Sefwi Wiawso District. The project is financed by the Social Fund of the Food Industry of Belgium. The objective of the project is to enhance a sustainable management of forest resources and promote food security and biodiversity conservation in the district and Ghana in general by promoting the farming of NTFPs.

The underlying assumption of the project is that the production of these products on a permanent basis will enormously help to create more sustainable employment and income-generation opportunities, enhance food security and the livelihoods of farmers, reduce the pressure on forest resources in the district and thereby ensure sustainable forest management. The project essentially focuses on **four** NTFPs: **grass-cutters, honey, mushrooms and snails**. These products constitute a very important source of food, nutrition and income for the population of the district. *After three years of implementation of the project, the results indicate that NTFP farming is positively impacting the district in economic, social and environmental terms.*

5.4 *Impacts of NTFPs Farming*

The results of a field survey on NTFPs farming conducted in the Sefwi wiawso District in 2006 indicate that these products generate tremendous benefits in terms of poverty reduction, food security and environmental management in the district. The main impacts of NTFPs are discussed briefly in the next sections of the paper.

The economic importance of NTFPs was determined using income, employment and household assets as indicators. The net income analysis (Cavendish, 2002; Marshall, 2005; Wong, 2000) was used to determine the economic importance of NTFPs farming in the district. Net income was defined as (gross income less expenses). The results indicated that NTFPs farming produced enormous benefits in terms of income generation and poverty reduction in the district.

The average annual income of NTFPs farmers in the district improved significantly since the introduction of NTFPs. About 71.7% of the farmers have moved above the Ghana upper poverty line established by the Ghana Extended Poverty (ESPS) and the Ghana Living Standard Survey (GLSS, 1998). Using the World Bank less than US$1:00 a day indicator reveals that 55.5% of the NTFPs farmers have moved out of poverty line since the introduction of NTFPs farming in the district. The acquisition of assets by the NTFPs farmers has also substantially increased in 2005 compared to three years ago. While the proportion of respondents with only one asset (*television or radio or fridge*) reduced in 2005, the number farmers that acquired more than one asset rose. For example, in 2003-2004 the number of respondents owning a television and radio increased from 22.0% to 33.1%, whiles those who acquired all basic assets rose from 3.5% to 13.8%. This is a clear indication that if farmers start reaping the full benefits of their farming activities in the years to come, the acquisition of assets, improvement of livelihoods and poverty reduction will significantly improve in the years.

5.4.1 Empowerment and Income Effects

The results of the study reveal that annual average incomes of NTFPs farmers in the district have improved significantly since the introduction of NTFPs. The study indicates that 71.7% of the farmers have moved above the Ghana upper poverty line established by the Ghana Extended Poverty (ESPS) and the Ghana Living Standard Survey (GLSS, 1998). Using the World Bank less than US$1:00 a day indicator reveals that 25.5% of the NTFPs farmers have moved above less than US$1:00 a day since the introduction of NTFPs farming in the district. The acquisition of assets by the NTFPs farmers has also substantially increased in 2005 compared to 2003. While the proportion of respondents with only one asset (*television or radio or fridge*) reduced in 2005, the number farmers that acquired more than one asset rose. For example, in 2003-2004, the number of respondents owning a television and radio increased from 22.0% to 33.1%, whiles those who acquired all the basic assets rose from 3.5% to 13.8%. This is a clear indication that if farmers start reaping the full benefits of their farming activities in the years to come, the acquisition of assets and livelihoods will significantly improve.

5.4.2 Education and Health

NTFPs farming has helped 23.5% of the farmers to put their children back in school as a result of improved income generated from NTFPs farming. They were compelled to take the children out of school because of inability to pay school fees. A cross-check conducted in five basic schools revealed that 15 children of NTFPs farmers were reinstated. Although, the study did not attempt a clinical analysis of the nutritional value of these products and their impact on human health, the results indicate that since the introduction of NTFPs farming in the district, the number of hospital visits by NTFPs farmers and their families has declined. Farmers who did not visit hospital increased from 33.8% in 2003 to 49.7% in 2005. The proportion of respondents who visited hospital once a year was 46.9% in 2003 but reduced to 33.1% in 2005. It is not very clear whether the improvement in the health of the respondents is a direct result of the nutritional value of the NTFPs. Further studies are currently being conducted to determine the nutritional value of these products and their impact on human health.

5.4.3 Food Security

Most of the sampled farmers are increasingly relying on NTFPs to meet their food and other needs. The field survey revealed that 86.7% of the respondents use NTFPs as food and medicines. These products include fruits and nuts, honey, grass-cutters, medicinal plants, mushrooms, and snails. Significant interrelationships between improved annual income generated from NTFPs farming, food security, education, health, and acquisition of assets were observed.

5.4.4 Environmental Benefits of NTFPs Farming

NTFPs farming has significantly benefited both the farmers and the environment of the district. An important aspect of NTFPs farming is its potential to shift farmers from slash and burn to more integrated farming methods. NTFPs farming in the district helped to move farmers from the devastating slash and burn method of farming to integrated agricultural practices. The number of farmers who have shifted from the devastating slash and burn method of farming to more sustainable ways of farming since the introduction of the NTFPs farming has increased significantly in the district. For example, the number of farmers practicing "zero burning" and integrated farming systems has increased from 5.5% in 2003 to 56.5 in 2005. Beekeeping is also facilitating a sustainable use of the forest; the installation of beehives in forest forces the farmers to desist from the usual slash and burn agricultural practice.

6. Challenges and Prospects of NTFPs Farming

Despite the fact that NTFPs farming in the Sefwi Wiawso District has had significant impacts on poverty reduction, sustainable forest management and livelihoods improvement, a number of factors continue to constrain the ability of farmers to exploit the full potential of NTFPs farming. Inadequate finance, limited training opportunities, passive community support, lack of inputs and farm construction materials, land tenure issues, and marketing constrains are the main challenges being faced by 88.3% of the NTFPs.

6.1 Policy Vacuum

The policy to guide the development, management, promotion and use of NTFPs in Ghana is relatively new. Until the promulgation of the Forest and Wildlife Policy in 1994, legislation on the forestry sector emphasized the supply of timber for the wood industry. NTFPs were not considered in forest management and planning. The current forestry policy has not adequately addressed issues of NTFPs.

6.2 Inadequate Finance

Provision of credit to farmers has always been an important factor in improving agricultural productivity and strengthening the rural economy in Ghana. Inadequate finance to start up NTFPs farming has been identified as the most critical economic challenge facing NTFPs farmers in the Sefwi Wiawso District and Ghana in general. Most farmers rely heavily on commercial banks for credits facilities but the rigid character of the banks is not friendly to the requirements of small-scale farmers. Most of the poor farmers also have no collateral to secure bank credits.

6.3 Lack of Inputs and Farm Construction Materials

The continuous decline of wood products for the construction of NTFPs farms calls for effective strategies in the areas of agro-forestry and re-afforestation. Wood is becoming very scarce and expensive in the Sefwi Wiawso District due to indiscriminate logging over the years. Consequently, prospective farmers find it extremely difficult to buy wood for the construction of bee hives, snails, mushrooms and grass-cutter rearing structures. Currently, it cost a farmer an average of $300-$350 to construct a beehive, or grass-cutter, mushrooms, and snails farm structure.

6.4   Lack of Awareness Creation

The potential contribution of NTFPs farming to improving, poverty reduction, livelihood and sustainable forest conservation has not been adequately promoted in the Sefwi Wiawso District and Ghana in general. The benefits of NTFPs farming have not been fully exploited, even though a big potential exists in rural areas, especially in the forest communities.

6.5   Limited Capacity Building Opportunities

Human and institutional capacity of NTFPs farmers and the various stakeholders involved in natural resource management, food security, and the promotion of NTFPs farming are not adequately strengthened to enable them to effective transfer of knowledge, skills, abilities and practices to the rural population. Many of the CSOs that are deeply involved in the development and promotion of NTFPs are also ill-equipped technically, financially and logistically.

6.6   Marketing Constraints

Most products like mushrooms get spoilt within a few days after harvesting. The deplorable state of feeder road network in the country, particularly the roads linking agricultural communities to market centres affects the marketing of NTFPs in Ghana. Absence of local food processing companies to process and can NTFPs in order to add value to the products also affects the marketing of the products both locally and internationally.

7. Recommendations for Enhancing the Development and Promotion of NTFPs

For NTFPs farming to be pivotal to poverty reduction, sustainable forest management and human development in Sefwi Wiawso District and Ghana in general, a number of appropriate strategies are recommended in this section. An effective implementation of these strategies by development policy-makers and actors will significantly facilitate the development and promotion of NTFPs, farming poverty reduction, conservation of forest resources and sustainable development in Ghana.

7.1 Formulation of a Visible National NTFPs Policy

As mentioned earlier, forest policy in Ghana has not adequately addressed issues of NTFPs. An appropriate policy framework for the sustainable promotion of NTFPs is necessary to help to ensure an effective development and promotion of NTFPs in Ghana. Such a policy will also strengthen agricultural research institutes and offer them the opportunities to develop and supply new breeding seeds and stock to NTFPs to farmers.

7.1.1 Prevention of Deforestation

Widespread deforestation in the Sefwi Wiawso District and Ghana in general, has led to the deterioration of the ecological system, particularly the marginal ecological zones. If the necessary efforts are not made to redress the situation, these problems will have enormous consequences on biodiversity loss and the livelihood of the population. NTFPs farming has the potential to help halt this decline and improve the livelihoods of the population. NTFPs farming, especially beekeeping provides an incentive to farmers to prevent the outbreak of bush fires. It is also an incentive for farmers to engage in agro-forestry and afforestation projects. It is therefore important that the government capitalises on the opportunity offered by NTFPs farming to educate the rural population to actively engage in NTFPs production reducing as a means of redressing poverty and protecting the environment.

7.1.2 Sustainable Forest Management

Traditional forest management, which focuses solely on timber exploitation, can undermine livelihoods of not only the rural forest communities but also sustainable forest management strategies. Therefore programmes and strategies formulated to effectively manage forests reserves in Ghana must pay greater attention to NTFPs farming. These products constitute an essential development alternative to the prevalent over-exploitation of forest resources, the devastating slash and burn method of agriculture and the indiscriminate and illegal logging of forests. Promoting NTFPs in Ghana will engender increased awareness on the application of poverty reduction, improvement of livelihoods appropriate and sustainable approaches to natural resources and forest management.

7.2 Provision of Financial Support to NTFPs Farmers

Lack of financial assistance to farmers to start up NTFPs farming has always been an important challenge to NTFPs farmers in Ghana. The rigid character of commercial banks also is not friendly to the requirements of small-scale NTFPs farmers. What is more, most of the poor farmers have no collateral to secure bank credits. It is therefore important that the Government introduces measures to strengthen the existing rural and agricultural development banks and encourage them to provide low interest loans to farmers to NTFPs farmers. In this way the pressure on forest resources will be reduced.

7.3 Supply of Construction Materials

One of the challenges of NTFPs farmers is the continuous decline of wood products for the construction of NTFPs farms calls for effective strategies in the areas of agro-forestry and re-afforestation. Wood is becoming very scarce and expensive in the Sefwi Wiawso District due to indiscriminate logging over the years. This has affected the cost of constructing materials necessary required setting up NTFPs farming. It is therefore recommended that new and more innovative and sustainable materials need to be developed for this purpose.

7.4 Need for an Effective Awareness Creation Programme on NTFPs

A vigorous promotion of NTFPs farming constitutes a very effective mechanism for popularizing the farming of NTFPs as an instrument for *poverty reduction* and ensuring *sustainable management of natural resources* in Ghana. Awareness creation of the nutritional, medicinal, and environmental benefits of NTFPs by Ministries of Agriculture, Health and Environment will significantly help to promote the farming of NTFPs in Ghana.

7.4.1 Vigorous Environmental Education

One important strategy for ensuring a sustainable management of NTFPs farming which can positively impact on poverty reduction is continuous public environmental education. It is recommended that vigorous environmental education be made part of natural resource management in the district.

7.5 Strengthening the Capacity of NTFPs Farmers and Relevant Stakeholders

Human and institutional capacity of NTFPs farmers and the various stakeholders involved in natural resource management, food security, and the promotion of NTFPs farming needs to be strengthened to enable them ensure an effective transfer of knowledge, skills, abilities and practices to the rural population. Many of these CSOs that are deeply involved in the development and promotion of NTFPs are ill-equipped technically, financially and logistically. Government should therefore subvent these organisations to enable them to effectively promote the development of NTFPs country-wide

7.5.1 Establising NTFPs Demonstration Farms

There is the need for MOFA to set up demonstration farms in rural communities to serve as centres of excellence for educating and training farmers and agricultural stakeholders on innovative and sustainable approaches to the production processing and marketing of their produce. In this way, appropriate agricultural technologies will be transferred to the population and thereby reduce poverty and improve food security on a sustainable basis.

7.5.2 Provision of Technical and Extension Services

One of the critical challenges faced by NTFPs farmers is inadequate technical and extension services: There is a need for the Department of Agricultural Extension of the Ministry of Agriculture (MoFA) to provide continuous technical, extension and monitoring services to NTFP farmers at the grassroots level. This will motivate and help farmers to adopt improved NTFPs farming practices and enable them to increase productivity and improve their livelihoods. There is also the need to provide farmers with the latest results of research on NTFPs farming techniques. MoFA should facilitate an effective linkage between the various research institutes and NTFPs farmers. This will enable farmers' problems to be brought to the relevant research institutes for investigation and redress.

7.6   Facilitating Processing and Marketing of NTFPs

Most products like mushrooms get spoilt within a few days after harvesting. To enhance the commercialization of NTFPs faming, it is important that processing facilities are provided. In addition, the government needs to improve the feeder road network in the country, particularly the roads linking agricultural communities to market centres. In order to add more value to NTFPs in Ghana, government also needs to facilitate the establishment of local local food processing companies to process and can NTFPs .

8. Conclusions

NTFPs farming activities provide an important opportunity for food security and poverty reduction for forest dependent communities. This paper has revealed that a significant number of NTFPs farmers and their households in the Sefwi Wiawso District in the Western Region of Ghana continue to supplement their incomes from NTFPs farming and are successful in satisfying their most important basic needs. For many farmers, engagement in NTFPs farming is due to the fact that the income they generate from traditional agricultural (mainly cocoa farming) is rather unreliable and woefully inadequate. NTFPs farming is therefore seen as an important component of rural survival and coping strategies; it forms part of their 'safety net'. A systematic, planned and a sustainable development and promotion of NTFPs in Ghana and Africa in general will enormously help to create all-year-round employment and income-generation opportunities, reduce poverty improve agricultural productivity, enhanced food security, poverty reduction and a sustainable of management of forest resources and biodiversity conservation.

9. Bibliography

Akrasi K. (2003). The State of the Forests in the Sefwi Wiawso District, paper presented at the Forest Resources Creation capacity building workshop, Sefwi Wiawso Pp 1-15

Andel, T.R. van (1998). Commercial exploitation of non-timber forest products in the Northwest District of Guyana. *Caribbean Journal of Agriculture and Natural Resources* Pp 15-28.

Andel T. R van (2000). Non-timber forest products of the North-West district of Guyana Part II A Field Tropenbos-Guyana guide Pp 1-281

Arnold, J.E.M. and Ruiz Perez, M. (1996). Framing the issues relating to non-timber forest products research. In: Ruiz Perez, M. and Arnold, J.E.M. (eds). Current issues in nontimber forest product research, CIFOR / ODA, Bogor, Indonesia, pp.1–18.

Arnold Michael, Ian Townson. (1998). Assessing the potential of forest product activities to contribute to rural incomes in Africa, Natural Resource Perspectives Number 37

Arnold, J.E.M. and M. Ruiz Perez. (1999). The Role of Non-timber Forest Products in Conservation and Development, in E Wollenberg & A Ingles (eds.), *Incomes from the forest: Methods for the development and conservation of forest products for local communities*. CIFOR / IUCN, Pp17 – 42.

Arnold, J.E.M. and Ruiz Perez, M. (1998). The role of nontimber forest products in conservation and development. In: Wollenberg, E. and Ingles, A. (eds.) Incomes from the forest: methods for the development and conservation of forest products for local communities, CIFOR / IUCN, Bogor, Indonesia, Pp17–42.

Arnold J.E.M. and P. Bird (1999). Forests and the Poverty-Environment Nexus, Prepared for the UNDP/EC Expert Workshop on Poverty and the Environment, Brussels, Pp 1-24

Arnold, J.E.M. and Ruiz Pérez, M. (2001). Can non-timber forest products match tropical conservation and development objectives? *Ecological Economics* 39: 437-447

Asante Alfred (2004) Assessment of Food Import and Food Aid against Support for Agricultural Development. The Case of Ghana, For Food and Agriculture Organization Regional Office, Accra, Ghana

Bhattarai, B., Ojha, H, Banjade, M.R. and Luintel, H. (2003). The effect of NTFP market expansion on sustainable local livelihoods– A case of Nepal, A discussion note. Forest Resources Studies and Action Team Pp 1-11

Boon E K, (2005) Combating Poverty in Africa, in Encyclopedia of Life Support Systems, UNESCO, Paris, France

Boon Emmanuel (2003). Environmental planning and management, paper presented at the Forest Resources Creation capacity building workshop for members of District Environmental Management Committee, Sefwi Wiawso, Ghana.

Chandrasekharan, C. (1995). Terminology, definition and classification of forest products other than Wood. Paper for the International Expert Consultation on Non-Wood Forest Products. 17-27 January, Yogyakrta, Indonesia. Pp 1-23

Food and Agriculture Organization of the United Nations (FAO) 1996. Rome Declaration on World Food Security and World Food Summit Plan of Action. World Food Summit 13-17 November 1996. Rome.

Food and Agriculture Organization of the United Nations (FAO) (2002). The state of food insecurity in the World 2001, FAO. Rome.

Food and Agriculture Organization of the United Nations (FAO) (2003). Trade Reforms and Food Security: Conceptualizing the Linkages, FAO, Rome

Fox, Jefferson. (1994). Introduction: Society and Non-Timber Forest Products in Asia. Society and Natural Resources. Vol.8, Pg. 189-192.

Ghana Statistical Services (2000), Ghana Living Standards Survey: Report on the Fourth Round (GLSS4), Accra, Ghana.

Ghana Government (2003), Ghana Poverty Reduction Strategy 2003–2005: An Agenda for Growth and Prosperity, Accra, Ghana.

Harry A. Sackey (2005), Poverty in Ghana from an Assets-based Perspective: An Application of Probit Technique, Accra Ghana

Malhotra, K.C. and Mark Poffenberger. (1989). *Forest Regeneration Through Community Participation*. West Bengal Forest Department. Calcutta, India.

Maxwell, S., Smith, M. (1992). Household food security; a conceptual review. *In* S. Maxwell & T.R. Frankenberger, eds. Household Food Security: Concepts, Indicators, Measurements: A Technical Review. New York and Rome: UNICEF and IFAD.

Nelson, W. (2007). Achieving Effective International Cooperation: Investinf in Environment for Development, The Case of SEA and the MDGs Experience in Ghana, Paper presented at EU Green Week, Brussels, Belgium w12-15 June 2007.

OECD (2001), The DAC Guidelines–Poverty Reduction, OECD Publications, Paris, P.31

Peters, C. M. (1996). The ecology and management of non timber forest resources. World Bank Technical Paper 322. Washington, D.C., USA.

Ruiz Perez, M. and Byron, N. (eds) (1999). A methodology to analyse divergent case studies of non-timber forest products and their development potential. *Forest Science,*
45, Pp 1–14.

Shiva M.P., Verma S.K. (2002). Approaches to Sustainable Forest Management and Biodiversity Conservation: With Pivotal Role of Non-timber Forest Products. Centre for Minor Forest Products, Valley Offset Printers, Dehra Dun, India. Pp 172

Sackey, H.A. (2005), Poverty in Ghana from an Assets-based Perspective: An Application of Probit Technique, Accra Ghana, Pp 1-32

United Nations Development Programme (UNDP) 2004. The Equator Initiative. *"Money Grows on Trees"*, Cameroon series 5

World Bank (2001) Poverty Trends and Voices of the Poor, The World Bank, Washington, DC

World Bank (2002), The Cost of Attaining the Millennium Development Goals, Washington, DC

Internet Sources

World Development Indicators 2004
http://www.worldbank.org/data/wdi2004/index.htm http://www.worldbank.org/data
(Accessed on 19/10/2006)

Ghana at a glance
http://www.worldbank.org/data/countrydata/aag/gha_aag.pdf (10/25/05)

(Accessed on 20/10/2006)

Prakash Loungani 2003. The Global War on Poverty:
http://www.imf.org/external/pubs/ft/fandd/2003/12/pdf/basics.pdf

(Accessed on 20/10/2006)

WSSD (2002)
http://www.un.org/esa/sustdev/documents/WSSD_POI_PD/English/WSSD_PlanImpl.pdf
(Accessed on 20/10/2006)

Louis, Kimberly, (2006).The Lament of a Nation: Ghana's SAP Experience.
http://www.ghanaweb.com/GhanaHomePage/features/artikel.php?ID=80453
(Accessed on 20/10/2006)

www.worlbank.org/poverty/mission/up1.htm

(Accessed on 15/11/2006)